HOME ROOFING REPAIRS

by James E. Brumbaugh

THEODORE AUDEL & CO.
a division of
HOWARD W. SAMS & CO., INC.
4300 West 62nd Street
Indianapolis, Indiana 46268

FIRST EDITION

FIRST PRINTING—1975

Copyright © 1975 by Howard W. Sams & Co., Inc., Indianapolis, Indiana 46268. Printed in the United States of America.

All rights reserved. Reproduction or use, without express permission, of editorial or pictorial content, in any manner, is prohibited. No patent liability is assumed with respect to the use of the information contained herein. While every precaution has been taken in the preparation of this book, the publisher assumes no responsibility for errors or omissions. Neither is any liability assumed for damages resulting from the use of the information contained herein.

International Standard Book Number: 0-672-23839-X

Preface

Every building erected must have some type of roof construction. The shape of the roof will naturally depend on the type of building, its usage, and the weather conditions of that particular area. Another important factor in roof construction is the material used—whether it is wood shingles, tile or slate, metal, built-up, or asphalt shingles.

Most well constructed asphalt shingle roofs should last from ten to twenty years, barring any severe weather conditions such as high wind or hail. Metal roofs should last many years if maintained properly with the proper paint or tar coating.

No matter what type of roof material is used, proper attention is needed to prevent leaks which can damage the wood construction and eventually the building interior. This book guides you through various steps in roof repair or replacement no matter what type of material is used. It explains various procedures in the installation of slate or tile roofs and also gives helpful hints on how to find leaks, as well as safety practices to follow in the repair of your own roof.

James E. Brumbaugh

CONTENTS

SECTION 1
Roofing Fundamentals and Safety Practices 7
A Word About Safety—Ladders and Scaffolds

SECTION 2
Roof Construction Details 16
Roof Leaks—Roof Flashing

SECTION 3
Asphalt Shingling 37
Staggering Shingles—Open Valleys—Covered Valleys—Shingles—Repairing an Asphalt Shingle Roof—Reroofing—Gutters and Downspouts—Roll Roofing—Roll Roofing Repairs

SECTION 4
Built-Up Roofing 61
Repairing Built-Up Roofing

SECTION 5
Wood Shingling 64
Repairing Wood Shingles

SECTION 6

Tile Roofing 69
   Repairing A Tile Roof

SECTION 7

Slate Roofing 74
   Repairing A Slate Roof

SECTION 8

Metal Roofing 78
   Repairing Metal Roofs

SECTION 1

Roofing Fundamentals and Safety Practices

A well constructed roof should last from ten to twenty years, depending on weather conditions and the materials used in its construction; however, *no* roof is expected to last forever. Most roofs will require a certain amount of maintenance and repair on an annual basis.

It is to your advantage to have an understanding of basic roofing fundamentals, even if you do not intend to do the work yourself, because such knowledge can save you money when you deal with professional roofers. You will be able to appraise the work better as it is being done.

If you feel that laying a new roof or repairing a roof will be too complicated for you to handle, then contact a *reputable* roofing firm. Many are listed in the Yellow Pages, but you should check with the local Better Business Bureau about your choice before signing any contracts or making any commitments. The Better Business Bureau may have a file of complaints against the firm you choose (being in the Yellow Pages is only a listing, *not* a seal of approval). *Never* allow a door-to-door salesman to talk you into a "cut-rate" job. Very few reputable roofing firms do business this way.

A WORD ABOUT SAFETY

Most people would rather not work on a roof, because of the dangers involved. The height and pitch of most roofs are certainly factors that should not be taken lightly. Falling from a roof can result in serious injury and even death.

Never go on a roof if you are afraid of heights. It is no way to cure this type of fear. Don't try to prove something to yourself or others. Risking your life is simply not worth it. If heights do not bother you, you should still carefully plan the job and follow a definite safety procedure. *All* professional roofers plan the job before they ever get on the roof. The job planning should include how the ladders and scaffolds are to be set up, how the materials are to be transferred to the roof, and where the safety rope and safety harness are to be positioned.

The clothing you wear should be loose enough to enable you to move freely. Never wear shoes with leather or composition soles even on the driest roofs. Tennis shoes offer the best footing. Bare feet will also give good footing, but there is always the possibility of stepping on a rusty nail or picking up a splinter.

Many roofing materials are brittle—particularly asbestos shingles, tile, and slate—and can be broken when stepped on. This problem can be avoided by distributing your weight more evenly over the surface of the roof. When working on a tile roof, for example, try to distribute the weight of each foot over two tiles at a time. Professional workers lay a 1 × 6 board on flat roofs and work from it. On pitched roofs, a ladder should be used.

A rope and safety harness is recommended on pitched roofs for additional safety. Both should be made from ⅜-inch nylon rope, because this type of rope offers the greatest tensile strength. The amount of rope you will need depends on the particular job. It should be long enough so that one end can be tied to a nearby tree, window, or door frame, or some other object that will hold your weight without slipping. It should also be long enough to enable you freedom of movement over the surface of the roof. The rope should be inserted through a three-foot length of garden hose which should be positioned at the point where the rope crosses the roof ridge (Fig. 1). This will protect the roof covering materials along the ridge from the pressure caused by your weight.

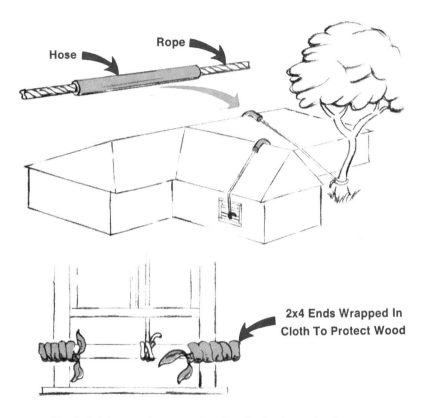

Fig. 1. Safety rope, hose guard, and methods of securing the rope.

A safety harness can be purchased through a mail-order house, or through a local building supply outlet. If you make your own safety harness, use ⅜-inch nylon rope and construct a safety harness similar to the one illustrated in Fig. 2.

Never work on a roof unless it is perfectly dry. Even dew moisture can make the roof slippery. Every precaution must be taken to ensure that you have secure footing.

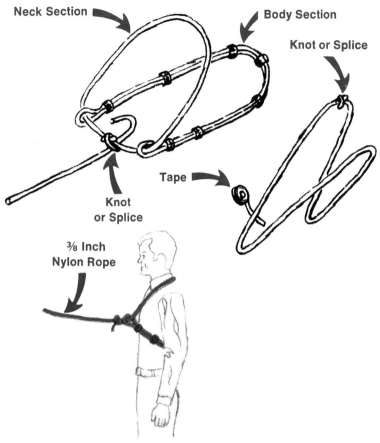

Fig. 2. Homemade safety harness. Rope for the neck and body sections should be cut to fit the worker's body measurements.

Don't get up on a roof if it appears that a storm is approaching. Chances are you won't be able to finish the job before it starts to rain anyway, and it will probably cause you to work faster than necessary and with less caution. There is also the danger of lightning. When a thunderstorm is nearby, the air will be charged with electricity. Lightning can be a threat without the clouds being directly overhead.

If there is a TV antenna on the roof, disconnect it. If you can't do it, have a professional come out and do it for you. Don't handle or go near power lines, TV antennas (even when disconnected), or other types of electrical devices or wiring.

LADDERS AND SCAFFOLDS

No roof repairs should be attempted without suitable ladders or scaffolds. Furthermore, they should be carefully

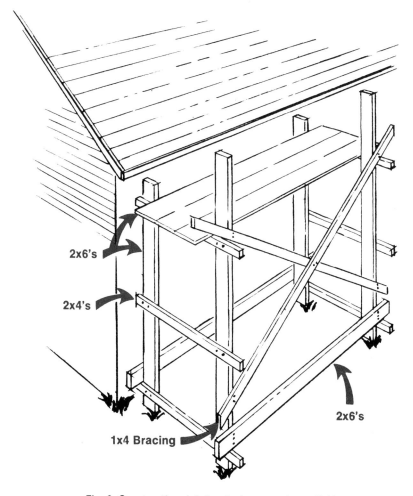

Fig. 3. Construction details of a homemade scaffold.

inspected for possible broken rungs or other defects, and *used properly.*

A scaffold can be used not only for getting up and down from a roof, but also for holding the materials nearby while you are working. This saves a lot of steps, because you do not have to climb down from the roof as often to get more supplies. The principal advantage of using a scaffold is that it does not lean against the gutter or eave, and therefore cannot cause any damage to these parts of the roof.

A scaffold can be purchased, built, or rented. Most towns of any size have Rent-All stores that will have one or more scaffolds on hand. Renting a scaffold is recommended over purchasing one, because the amount of time spent in roofing a structure is relatively short and any extensive reroofing should not be necessary for a number of years. Don't attempt to build a scaffold unless you are a good carpenter. A poorly built scaffold is worse than no scaffold at all. Scaffold construction details are shown in Fig. 3. The cross bracing is particularly important for safety. *All* handmade scaffolds should have cross braces.

Any ladder you use for access to the roof should extend far enough above the eave to allow you to step directly onto the roof. *Never* use a ladder so short that you must step over the top of it. This can be dangerous. Check the ladder over first for broken or loose rungs, or splits in the rails. If the ladder is defective, do *not* use it. Never use a painted ladder, because the paint could be concealing a cracked rung.

A ladder should be raised to the roof by "walking" it (Fig. 4). Place the bottom end of the ladder against the wall, and raise it until it is in a vertical position. Lift the bottom of the ladder and move it away from the wall a distance approximately one fourth its length. The ladder should be positioned so that it is absolutely straight (i.e. not leaning to one side or the other). The bottom of the ladder should be firmly placed so that the bottom of both rails rest solidly and evenly on the ground. If the ladder wobbles when you climb it, get down immediately and reposition it until it is stable. When raising a ladder, do *not* slam it against a gutter. It is very easy to damage a gutter in this way.

Never lean from a ladder, because unevenly placed weight may cause it to fall. Try to keep your hips between the two rails as you climb the ladder.

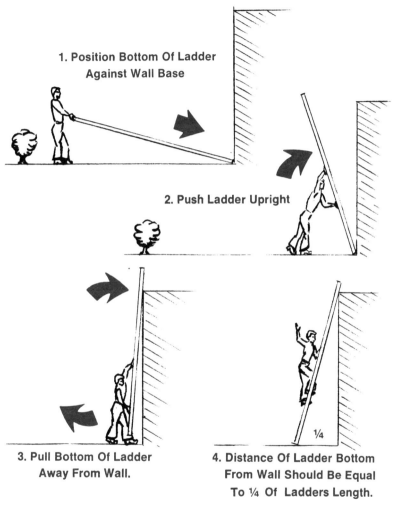

1. Position Bottom Of Ladder Against Wall Base
2. Push Ladder Upright
3. Pull Bottom Of Ladder Away From Wall.
4. Distance Of Ladder Bottom From Wall Should Be Equal To ¼ Of Ladders Length.

Fig. 4. Walking a ladder.

On a pitched roof, a ladder can be used for support by hooking it over the roof ridge as shown in Fig. 5. Ladder hooks can be homemade or purchased.

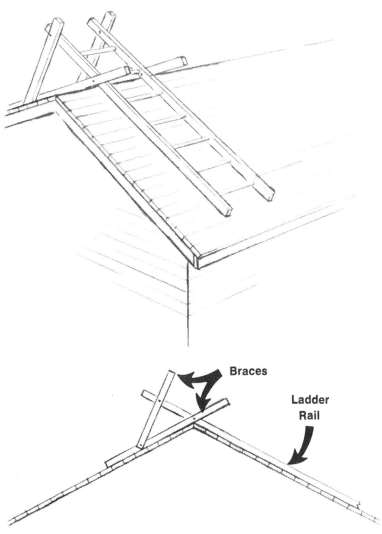

Fig. 5. Ladder with braces or supports attached to the rails. The braces must be set to the angle of the roof.

A one-board or "chicken ladder" (Fig. 6) is recommended for use on roofs of tile, slate, or other types of brittle and easily breakable roofing materials, because it spreads your weight more effectively. These are homemade ladders consisting of 1 × 10 or 1 × 12 board to which 1 × 2 pieces

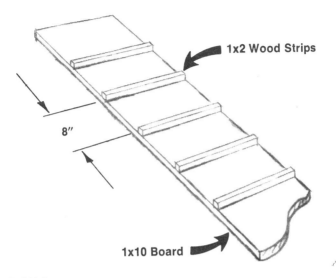

Fig. 6. Chicken ladder. Wood braces can be attached at one end to hook over the roof ridge.

of wood are nailed. A "chicken ladder" should also be hooked over the roof ridge.

Metal brackets for supporting a 2 × 4 can be purchased through most local building supply houses. The brackets are nailed to the roof and the 2 × 4 board provides footing on roofs with a steep pitch.

SECTION 2

Roof Construction Details

The construction details of a typical roof are shown in Fig. 7. Familiarize yourself with these details, because you

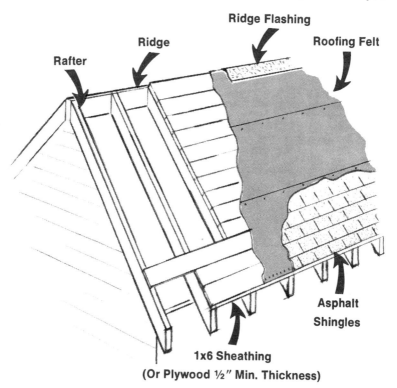

Fig. 7. Roof construction details.

will be confronted with the terminologly involved here throughout the book.

The *ridge* is usually stock about one inch thick inserted between the rafters at the top of the roof, the ridge should be wide enough to receive the whole depth of each rafter (Fig. 8).

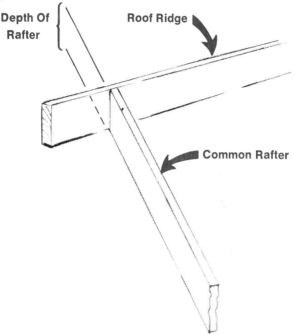

Fig. 8. Depth of rafter.

Rafters are used to support the roof surface. A *common rafter* extends all the way from the ridge to wall plate, and is not connected to any other rafter. Each rafter also extends a short distance beyond the wall plate. This is referred to as *projection* or *overhang*. A *jack rafter* is shorter than a common rafter, because it is connected at the upper end to either a *hip rafter* or *valley rafter*. A *cripple rafter,* on the other hand, is connected at both ends to a hip or valley rafter. The various types of rafters are illustrated in Fig. 9.

A *hip* or *hip roof* is one which has a sloping surface from each wall of the structure to the ridge. A *minor roof is* any

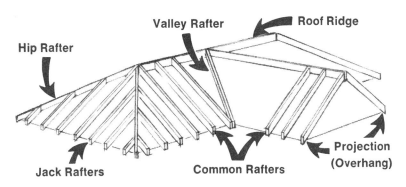

Fig. 9. Various types of rafters.

roof extending out from the main roof. It is connected to the main roof by a valley rafter. A *valley* is the portion of the roof formed by the meeting of a minor roof and the main roof (Fig. 10).

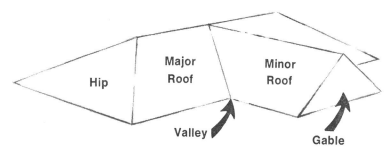

Fig. 10. Major roof, minor roof, hip, gable, and valley. A major roof with a hip (instead of a gable) is referred to as a *Hip Roof*.

The rafters are covered with *sheathing* which provides a surface for the roofing materials (Fig. 11). Sheathing generally consists of 1 × 6 boards or ⅝-inch *plyscord*. Other thicknesses of plyscord sheathing are also used.

The sheathing is generally covered with a base layer of roofing felt which is overlapped several inches. The amount of overlap and exposure is usually recommended by the manufacturer.

The *run* is a horizontal line extending from the fascia line at the outer edge of the rafter to a plumb line extending

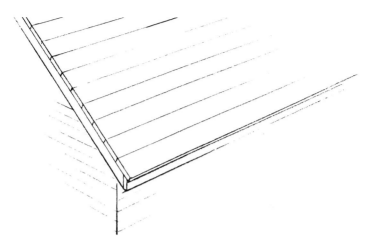

Fig. 11. 1 x 6 roof sheathing. On wood shingle and shade roofs the boards are spaced 2 to 3 inches apart.

down from the center of the ridge. The *fascia* or *fascia line* is the inside of the fascia forming the cornice (Fig. 12). A

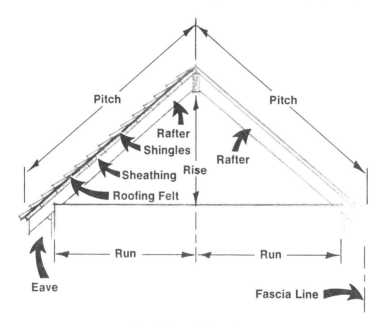

Fig. 12. Run, pitch and rise.

gable is the triangular area of wall at the end of a gable roof. Those portions of the roof that overhang the walls of a structure are the *eaves* (Fig. 12).

The *pitch* of a roof is the slant of the rafters expressed in inches to the foot. Pitches are referred to as ¼, ⅓, or ½ pitch as illustrated in Fig. 13. A ¼ pitch roof is one which will rise 6 inches for each foot of run. A ⅓ pitch rises 8 inches for each foot of run, and a ½ pitch rises 12 inches for each foot of run.

The pitch of a roof should be taken into consideration when laying a *new* roof, because it will determine the type

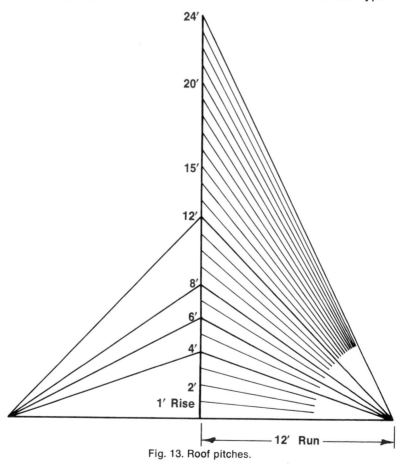

Fig. 13. Roof pitches.

of roof covering material used in many cases. For example, asphalt shingles can be used on roofs with a pitch of 4 inches or more, but require special application when the pitch is less than 4 inches. The special application usually requires the cementing down of each shingle with roofing cement. A salesman at the local building supply house should be consulted and his recommendations followed.

The roofing materials should be properly stored when not being used. Containers of asphalt roofing cement should be stored on end and level. Roll roofing should also be stored on end. Keep all materials in a heated and dry area, but far enough away from any heat sources to avoid a fire. Roofing materials are highly flammable.

Weather extremes can result in poor working conditions. The best temperature range is between 50°F and 80°F. If the temperature is below 50°F., keep the roofing materials —particularly asphalt shingles and asphalt roofing cement— in a dry, warm place before using them.

The roof sheathing provides the base for the roof covering materials. Two or more layers of a special asphalt impregnated roofing paper are laid over the sheathing to produce a waterproof cover. Finish roofing materials, such as shingles, tiles, or slate, are laid over the roofing paper. The roofing materials are laid in courses beginning at the eaves and working toward the ride on pitched roofs, or from one end of a flat roof to the opposite one. Each course overlaps the course laid before it. This overlapping of the courses enables the roof to shed water more easily without leaking through the sheathing, but it can complicate repairs. Minor repairs will necessarily constitute interruptions in this overlapping; therefore, you should examine how these materials are overlapped before you begin work.

ROOF LEAKS

Roofs are subject to all kinds of weather conditions and will sooner or later develop one or more leaks. The most common types of leaks in a roof are traced to one or more of the following causes:

1. Warped, corroded, or cracked flashing around chimneys, vents, and other structural interruptions in the roof surface.
2. Broken, loose, or missing shingles, tiles, or other types of roof covering materials.
3. Rusty, loose, or missing nails used to secure the covering materials to the roof.
4. Dried out and cracked roofing compound used to seal seams in roof covering materials.
5. Blocked or damaged gutters and downspouts.
6. Rotted or cracked roof sheathing board.
7. Ice dams forming along roof eaves.

Tracing Roof Leaks

Most roof leaks are difficult to trace because the water almost never collects directly under the point at which it enters. It will frequently run a considerable distance down a joist (rafter) or some other part of the structure before dropping and forming a puddle.

The best time to trace a leak is during the day when there is sufficient light. Find the point at which the water has been collecting and then look for water stains along the joints above it. These stains will usually lead you to the point at which the water has been entering. Sometimes you will be able to see a small pinhole of light. If you can locate the leak, shove a straight piece of wire through the hole to indicate its position from the outside. If you cannot find the point where the water is entering, calculate its approximate location by measuring the distance from the end of the joist to the point at which the stains end. Add to this measurement the length of the roof overhang, and this will be the approximate distance of the leak from the edge of the roof.

If it is raining, trace the flow of water up along the joist to its approximate entry point. Repairs are impossible until the rain stops, but you can mark the point at which the water enters and place a pail where the puddle is forming. To ensure that the dripping water does not change its position on the joist, run a string or wire from the joist down to the pail (Fig. 14). After you have located the leak from the inside (or at least determined its approximate location), go onto the roof and try to find the cause of the leak.

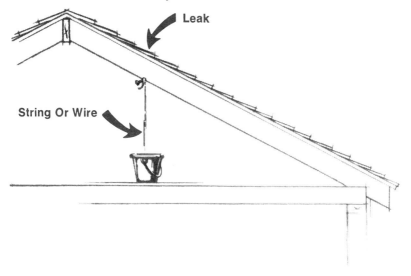

Fig. 14. Marking a leak with a string or wire.

Temporary Repairs

A temporary repair is any kind of repair that will stop or effectively contain the water leaking through the roof on a temporary basis until permanent repairs can be made. Repairing a leak is *not* recommended when the weather is bad. A wet roof offers very slippery footing and could be dangerous; therefore, you should spend as little time up there as necessary. The leak should be stopped with a temporary repair until it is possible to get up on the roof and make a permanent one. Once you have located the leak, you can cover it from the outside with a large piece of waterproof material (canvas, plastic, etc.).

Temporary repairs from the *inside* depend on where the leak is found. If it is a small one in the roof sheathing located away from a joist (rafter), ridge board or wall plate, a square of plywood and a 2 × 4 can be used (Fig. 15). Dry off the area and coat it with roofing cement. Coat one side of the

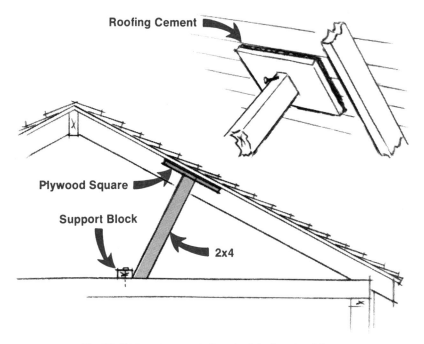

Fig. 15. Temporary repair for a leak in the sheathing.

piece of plywood with roofing cement and press it down over the area. The 2 × 4 should be precut so that it forms a tight fit between the plywood and the floor.

Small leaks around the edges of rafters, the ridge board, and wall plate can be stopped with roofing cement or putty, but this will not last very long.

Buckets and pails are necessary to catch the water for larger leaks, although the waterproof cover on the outside of the roof will, in most cases, reduce the amount of leaking water to a minimum.

ROOF FLASHING

Flashing is a material used in roofing to produce a watertight joint. It is frequently used where the roof joins a chimney, vent pipe, skylight, or some other vertical surface.

It is also used in valleys and along ridges, and less frequently along an eave or gable end. Typical flashing applications are illustrated in Fig. 16.

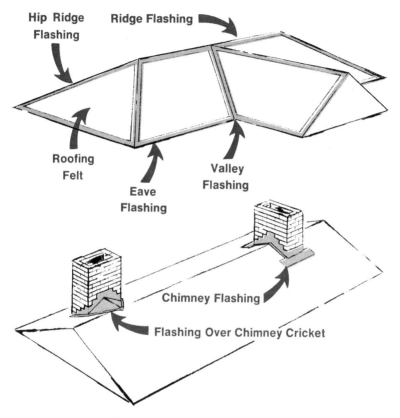

Fig. 16. Typical flashing applications.

Flashing is available in rolls of aluminum, copper, and galvanized iron, ranging in widths from 12 to 18 inches. Aluminum and copper are the more commonly used types of metal flashing. Aluminum flashing is becoming more popular, because it is less expensive than copper. Neither aluminum nor copper flashing require a special finish to protect the metal against corrosion. This is not the case with galvanized iron. Unless it is covered with a protective finish, it will rust and corrode. Even with the protective finish, a scratch or damage to the flashing can cause the metal to deteriorate.

Do not use ordinary nails to fasten the metal flashing to the roof. These will rust. Aluminum nails should be used with aluminum flashing; copper nails with copper flashing. Follow the manufacturer's recommendations for the type and size nail to use.

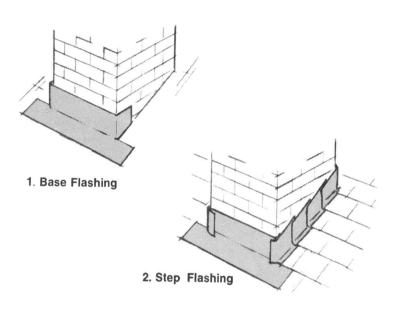

1. Base Flashing

2. Step Flashing

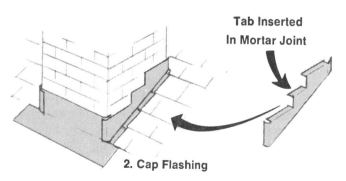

Tab Inserted In Mortar Joint

2. Cap Flashing

Fig. 17. Applying flashing to a chimney in three basic sections.

Both plastic and mineral-surfaced roll roofing are also used as flashing materials. Their principal advantage is that they are cheaper than metal flashing.

Flashing Around Chimneys and Vents

The joint between a chimney and the roof should always be protected with flashing, because this is probably the the most common source of roof leaks. Consequently, great care must be taken when applying the flashing to these areas to ensure a tight and long lasting fit.

The flashing around a chimney is sometimes built up from a number of different sections (Fig. 17). The *base* section is applied first and laid along the base or bottom of the chimney. It is frequently laid over the shingles. The flashing laid up the slope of the roof along the sides of the chimney is cut to the size of each course. This is sometimes referred to as *step* flashing. Step flashing is also applied to the back of the chimney, but the upper sections of this flashing (i.e. the section laid against the chimney) is cut long enough to extend several inches above the chimney cricket (see *Chimney Cricket*). After the base and step flashing are applied, a

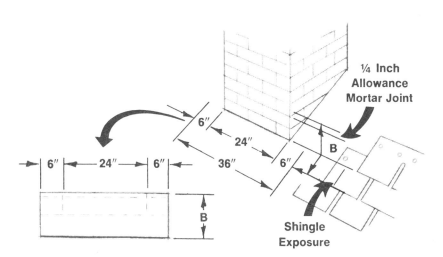

Fig. 18. Cutting out the base section.

cap or *counter* flashing is laid over them. This is a one piece section of flashing cut to size for each side of the chimney. The top edge of the counter flashing is bent (usually about ½ inch) and inserted in a mortar joint in the chimney and sealed with brick mortar.

The procedure for laying flashing around a chimney is illustrated in Figs. 18 and 22, and may be outlined as follows:

1. After the last course of shingles has reached the base of the chimney, take the measurements for the base flashing.
2. Cut out a rectangular piece of flashing material. Its length should equal the width of the chimney *plus* 12 inches. Its width will equal the combined distances from the base line of the chimney to a point equal to the exposure allowed for the shingles, and from the base line to ¼ inch *above* the mortar joint of the first or second course of chimney bricks (Fig. 18).

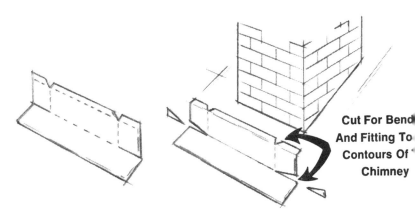

Fig. 19. Initial cuts in the flashing.

Fig. 20. Fitting the flashing to the contours of the chimney.

3. Bend the piece of flashing so that the bottom half lies flat against the roof and extends out to the middle of the last course of shingles, and the top half lies flat against the chimney. Position the flashing against

the chimney so that exactly 6 inches of flashing extends on either side.

4. Cut 6 inches in from each side of the flashing along the bend. Make two additional ¼ inch cuts in from the top of the flashing where it will bend around the chimney (Fig. 19).

5. Bend the top half of the base flashing to fit the contours of the chimney, and cut each side to fit the slope of the roof. Bend back ¼ inch of the top of this section to form a lip to be inserted in the mortar joint. Remove ¼ inch of mortar from the joint (Fig. 20).

6. Cover the bottom of the base flashing with asphalt roofing cement and nail it to the roof. Insert the ¼ inch lip in the mortar joint and seal it with brick mortar (Fig. 21).

Fig. 21. Mortar groove.

7. Cut and apply a separate section of step flashing for each course of shingles laid up the roof slope. Each section of step flashing is coated with asphalt roofing cement and nailed to the roof as shown in Fig. 22.

8. The cap or counter flashing is designed to cover both the base and step flashing. It is cut to size in one section for each side of the chimney (Fig. 17). It is bonded to the other flashing, and the top edge is bent to fit into a mortar joint.

9. The base flashing is not covered with shingles. The step flashing is covered with shingles up to the edge of the chimney.

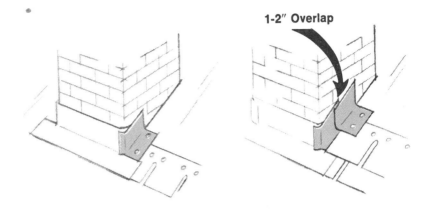

Fig 22. Applying step flashing.

Another method of applying flashing to a chimney involves applying it directly to the roofing felt and chimney *before* the shingles are laid (Fig. 23). In this method, the step flashing is cut much wider than those described in Step 7 above. The shingles are laid up to the edge of the chimney and bonded to the flashing with asphalt roofing cement.

Cap or counter flashing is not always used on chimneys. Some roofers will create a lip on the upper edge of each step flashing and insert the lip in a mortar joint.

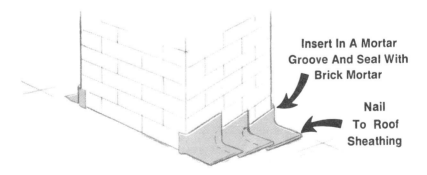

Fig. 23. Alternate method of applying flashing to a chimney.

Many roofers will further waterproof the flashing by applying asphalt roofing cement to all joints and along all edges (Fig. 24).

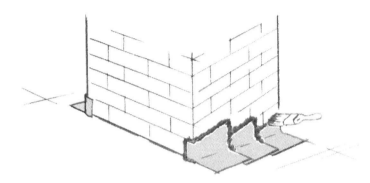

Fig. 24. Applying roofing cement to edges of flashing.

The flashing around vents and soil stacks is relatively uncomplicated to apply. Manufacturers often provide flared skirts and flanges (or shields) that function as flashing (Fig. 25). When these are provided, follow the manufacturer's instructions for attaching them to the roof.

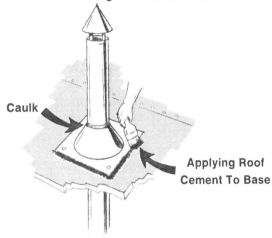

Fig. 25. Flashing around roof vent.

Chimney Crickets

A *chimney cricket* (or *saddle*) is a built up area on pitched roofs behind the chimney (Fig. 26). It is designed in the shape of a small false roof to throw off water that would

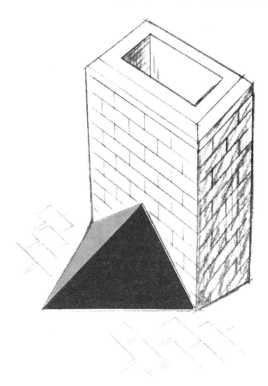

Fig. 26. Chimney cricket (or saddle).

otherwise collect behind the chimney and possibly cause leaks. It also prevents the buildup of snow which can result in the same problem. The construction details of a typical chimney cricket are shown in Fig. 27.

The top of the cricket is commonly located at a height equal to one half the width of the chimney. For example, if the chimney is 30 inches wide, the top of the cricket will be located 15 inches up from the point at which the chimney meets the roof (Fig. 27).

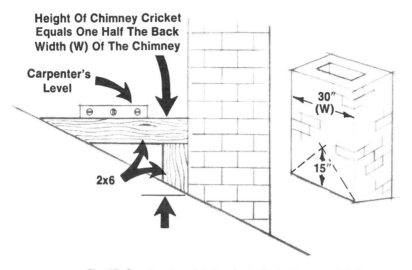

Fig. 27. Construction details of a typical chimney cricket.

Cut a piece of flashing to cover the cricket and allow an overlap of 4 or 5 inches on the chimney and roof (Fig. 28). Nail the flashing to the roof and cricket sheathing. Cover the nail heads with a dab of asphalt roofing cement. If the cricket is a large one, you may have to construct the flashing from two or more sections soldered together.

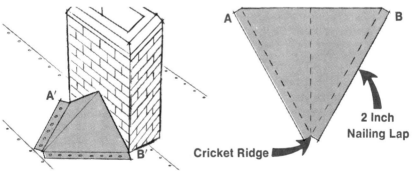

1. Coat Bottom Of Cricket Nailing Border With Roofing Cement
2. Nail Border Of Cricket To The Roof So That Points A And B Meet At The Outer Edge Of The Chimney (Points A' And B')

Fig. 28. Chimney cricket flashing.

Dormer Flashing

Step flashing should be applied up the slope of the roof against the sides of a dormer (Fig. 29). The procedure is the same as the one described for applying step flashing to chimneys.

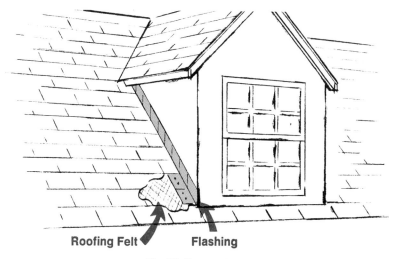

Fig. 29. Dormer flashing.

Valley Flashing

Valleys are formed when two roof sections, or a wall and a roof section, come together at an angle. These valleys will frequently extend from a roof ridge to the eave. Because of this, many builders substitute mineral-surfaced roll roofing as a flashing material rather than use the more expensive metal flashing. The major disadvantage of using roll roofing for flashing is its tendency to expand or contract during temperature extremes. The expansion or contraction of the flashing can cause the roofing cement along the edges of the flashing to crack and possibly leak.

The procedure for using mineral-surfaced roll roofing as flashing in a roof valley is illustrated in Fig. 30 and outlined as follows:

1. Measure off enough material from the roll to cover the entire length of the valley.

2. Strike a chalk line down the middle of this strip and cut it in half.
3. Trim the top of the strip to fit the angle of the roof ridge, and trim the bottom to fit the angle of the eave.
4. Take this half section and nail it in the valley with the mineral surface facing down. Use large head

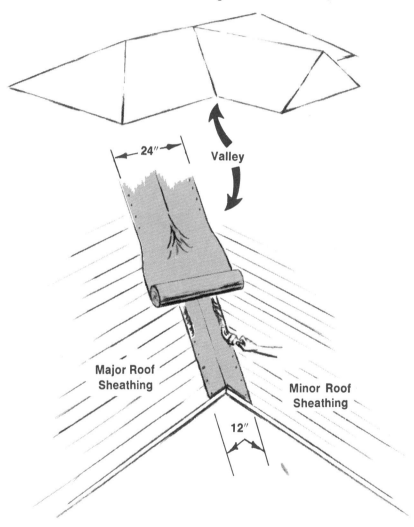

Fig. 30. Applying valley flashing.

roofing nails spaced 6 inches apart and 1 inch in from the outer edges of the strip.
5. Cut off another strip of material from the roll the same length as the first one, but do not cut it down the middle. Trim the top and bottom of the second strip to fit the roof angles.
6. Nail the second length of roll roofing down over the first one, but with the mineral-surfaced side facing up. Use the same nailing pattern described in Step 4 above.
7. For additional protection from leaks, seal the edge of the flashing with roofing cement.

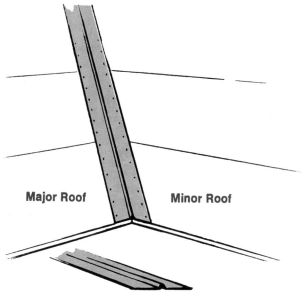

Fig. 31. Metal flashing with inverted vee.

Metal flashing (copper, aluminum, etc.) is also available in rolls, ranging up to 18 inches wide. Only one strip of metal flashing is used in a valley. The nails should not be positioned closer than 6 inches apart.

Some metal flashing is available ready made with an inverted vee running down to middle of its length to facilitate drainage (Fig. 31).

SECTION 3

Asphalt Shingling

Asphalt shingles are the most common type of roofing material used on pitched roofs. They are lightweight, comparatively inexpensive, and available in many different colors, sizes, and shapes.

Asphalt shingles are sold in the form of individual shingles or as strips of shingles joined together in two, three, or four tab units. The strips are 36-inches long. Their width will vary from slightly over 11 inches to 12 inches wide, depending upon the style and manufacturer. Lock-type asphalt shingles are also available in a number of different styles. Sizes will vary (Fig. 32).

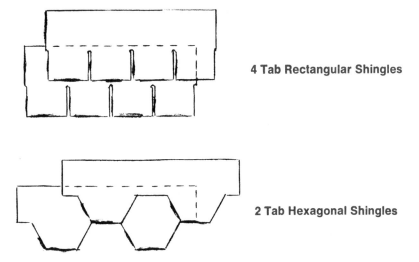

Fig. 32. Typical examples of asphalt shingle strips.

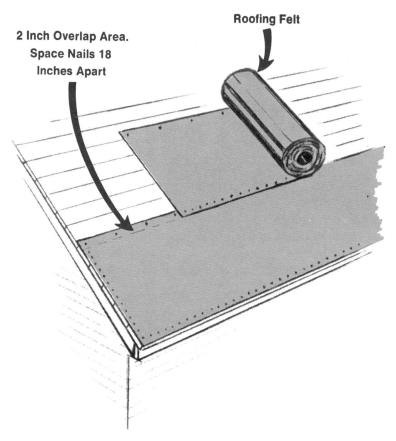

Fig. 33. Applying roofing felt. Nails are spaced 2 inches apart along the bottom edge of each course of roofing felt and the edges of the roof. Each course overlaps by at least 2 inches. Nails in the overlap area are spaced 18 inches apart. Overlap area is covered with roofing cement before the next course of roofing felt is laid. All nails are placed 1 inch in from the edge of the roofing felt.

Before laying a *new* roof with asphalt shingles, you must first calculate the total area to be covered. A high school plane geometry text will provide you with the formulas necessary to calculate the square footage of surface area. Many roofs form uncomplicated rectangulars or squares. Unfortunately, hip roofs, dormer roofs, and other types of minor roofs present more complicated surfaces. In any event,

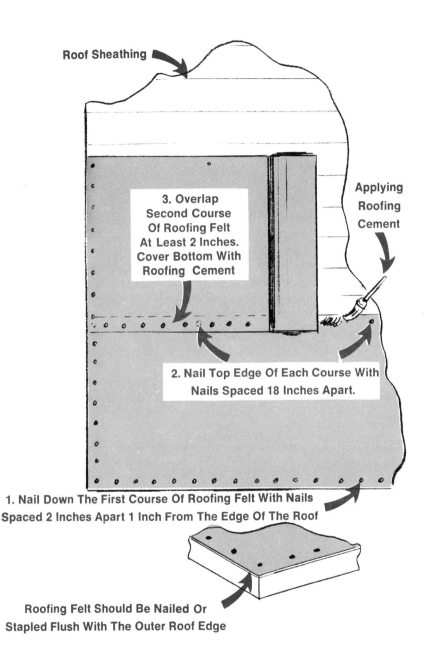

Fig. 34. Applying roofing felt.

the total square foot area of the roof covering the structure must be calculated.

When you are satisfied with your calculations, go to your local building supply dealer and purchase the shingles. Large quantities of shingles are sold in units (squares) that will cover 100 square feet of roof area. Each square will contain 3 bundles of asphalt shingles. You will also need nails. The type and number of nails is generally recommended by the shingle manufacturer. The manufacturer will also recommend the amount of exposure (usually about 5 inches) for the shingles and instructions for nailing.

Additional supplies for asphalt shingling include enough roofing felt or paper to cover the roof and serve as a base for the shingles; a suitable flashing material; and enough roofing cement to do the job.

Before asphalt shingles are applied, the entire roof must be covered with asphalt impregnated roofing felt. The roofing felt is stapled or nailed to the sheathing with each course overlapping the preceding one by approximately 2-3 inches (Figs. 33-34).

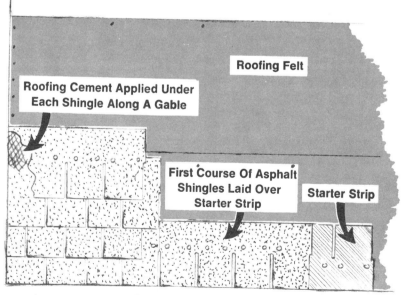

Fig. 35. Applying asphalt shingles.

When used, flashing along eaves and gable ends should be applied after the roofing felt has been laid. The flashing around chimneys, vents, and dormers, and along the roof ridge can either be applied before the shingles are laid, or when the asphalt shingles reach those points on the roof.

Many professional roofers begin by laying a starter strip of asphalt shingles along the roof eave. The shingles in the starter strip are laid face down with their top edge flush with the eave or flashing (Fig. 35). The first course of asphalt shingles is laid directly over the starter strip and nailed to the sheathing.

A chalk line is used to establish the position of successive courses of asphalt shingles. Each course is laid so that there is approximately 5 inches of exposure and the tabs are staggered (Fig. 36). The shingle manufacturer will generally recommend the type of nail and nailing pattern to use.

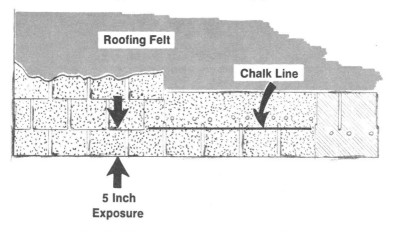

Fig. 36. Shingle exposure and use of chalk line.

STAGGERING SHINGLES

Each course of shingles should be laid so that the breaks between individual shingles (or tabs) are staggered. This can be done by beginning at the center of the roof and working toward the gables (or hips) and then cutting off the excess shingle to conform to the edge of the roof when it meets the gable. Another method is to begin at one end of the roof and

41

work toward the other. Both methods involve trimming the shingles to conform to the roof edge (Fig. 37).

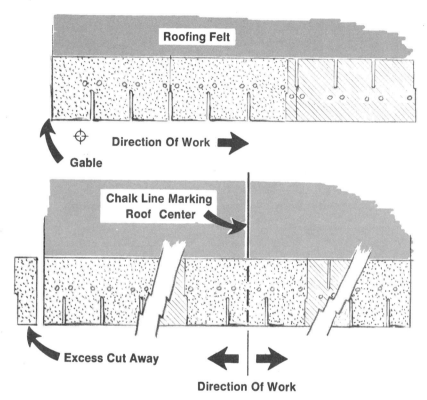

Fig. 37. Direction of work.

OPEN VALLEYS

On some roofs, the valleys are left exposed. The shingles are cut at an angle parallel to the roof joint, but overlapping the flashing by 2 or 3 inches. A chalk line should be struck the length of the flashing 2 or 3 inches in from each edge (Fig. 38). The shingles that form the border with the flashing should be coated with asphalt roofing cement and nailed to the roof (Fig. 39). When the flashing is left exposed, it should be the same color as the shingles.

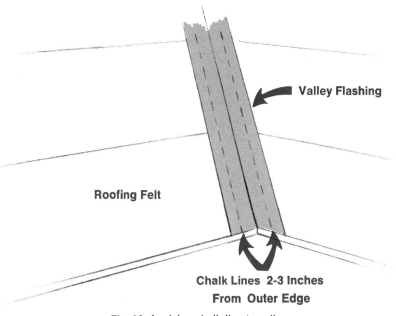

Fig. 38. Applying chalk line to valley.

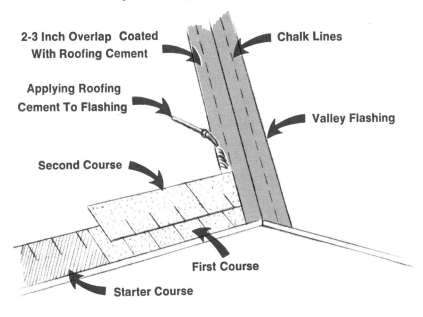

Fig. 39. Open valley.

COVERED VALLEYS

Some workers prefer to cover the roof valleys with shingles rather than to leave them open. There are two ways to do this. One method is to cut the shingles on both the main and minor roofs so that they form a close fit. The valley is lined with a double course of roofing felt first, and the shingles in the valley are fastened to the roof with nails and asphalt roofing cement. This procedure is illustrated in Fig. 40.

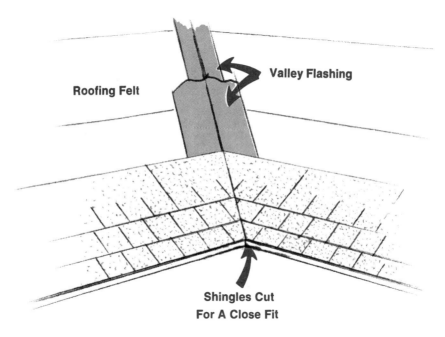

Fig. 40. Closed valley.

The second method of covering a roof valley consists of shingling over it in overlapping courses. This is possible with asphalt shingles, because they are flexible and can be bent to conform to the angle of the valley. Care must be taken, however to bend them carefully and not break them. Because these overlapping courses of shingles do not lie flat against the surface of the valley, you should apply valley flashing beneath them to protect the roof joint from leaks (see Valley Flashing). The method for shingling over a valley with overlapping courses is illustrated in Fig. 41.

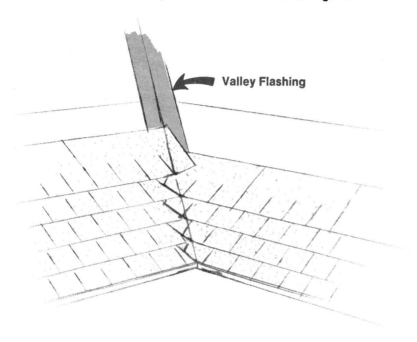

Fig. 41. Method of overlapping a valley with flashing.

SHINGLING HIP ROOFS

A hip roof is one which slopes upward from each wall toward the ridge. The degree of slope will depend on the pitch of the hip rafters, which extend from the outside corners of the structure to the ridge or a common rafter (Fig. 9).

The recommended method of shingling a hip roof is to lay each course of shingles completely around the entire roof (Fig. 42). Cut and fit the shingles of each course around each hip ridge. When all the courses have been laid, cover each hip ridge with a course of shingles. Begin at the eave, and lay the course of hip ridge shingles in the direction of the main roof ridge. The hip ridge shingles must be applied

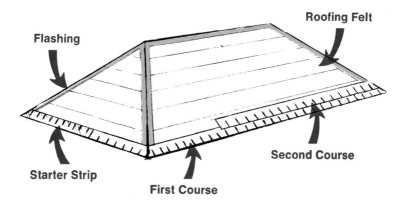

Fig. 42. Shingling a hip roof.

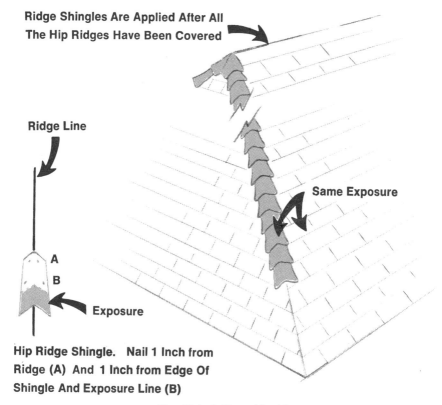

Fig. 43. Installing a hip ridge.

before the shingles on the main roof ridge. Four nails are used with each hip ridge shingle. Two of the nails should be very close to the hip ridge. The other two are approximately 1 inch from the outside edge of the shingle. (Fig. 43).

RIDGE SHINGLES

Ridge shingles are used to cover the ridge line of the main roof (Fig. 44). The ridge shingles on a main roof are started at *both* ends. The work moves toward the center of the roof where both courses meet. Four nails are used to attach each ridge shingle. Ridge shingles can be made by cutting ordinary shingles to the size required, or specially made shingles for this purpose can be purchased.

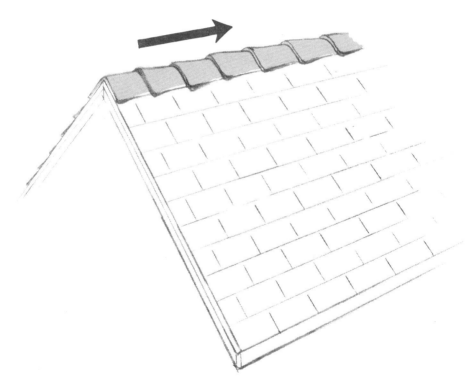

Fig. 44. Ridge shingles along main roof ridge.

REPAIRING AN ASPHALT SHINGLE ROOF

One problem encountered when repairing an asphalt shingle roof is getting new shingles that will match the old ones. There are quite a few different manufacturers, and it is not uncommon for them to discontinue a color or style. If the asphalt shingles on your roof represent a discontinued line, you will have to find replacements that match the old ones in color and style as closely as possible; otherwise, the new shingles will provide a sharp contrast with the older

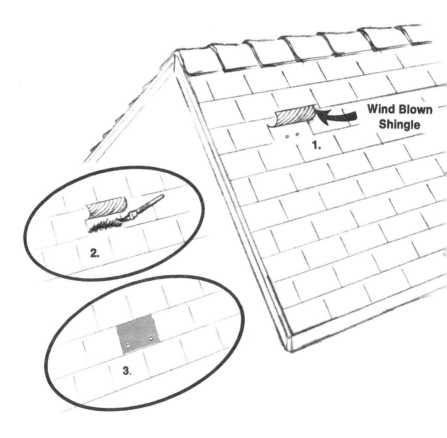

Fig. 45. Repairing a windblown shingle.

ones. Take a piece of the damaged shingles into your local building supply house and try to match it in weight, color, and style. Try to get the overall dimensions of the shingle (e.g. 36 inches × 12 inches, 12 inches × 12 inches, etc.) so that you can also match sizes.

Strong winds will sometimes cause a shingle or shingle tab to bend and peel up from the roof. If the shingle has not been broken, it can be repaired by spreading asphalt roofing cement on the bottom of the shingle and pressing it down. If the shingle won't lie flat, nail it at the edge and cover the nail head with a dab of roofing cement (Fig. 45).

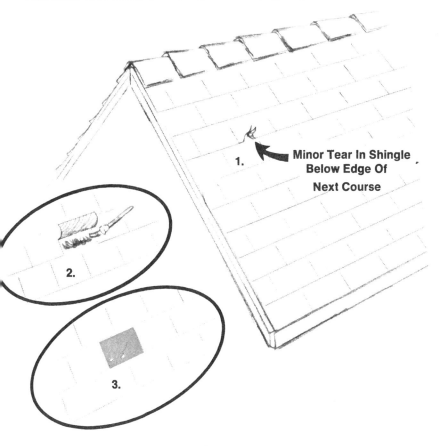

Fig. 46. Repairing a minor tear.

Small tears can be mended in the same way if they do not extend up under the overlapping shingles of the next higher course. Coat the bottom of the torn shingle with roofing cement, firmly press the two sections together and down with roofing cement (Fig. 46).

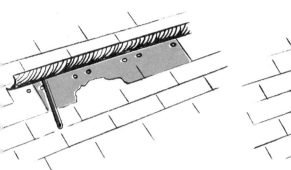

1. Bend Up Next Course And Cut Nails From Damaged Shingle.

2. Remove Shingle

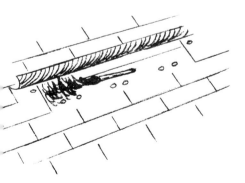

3. Coat Area With Roofing Cement, Insert New Shingle, And Nail In Place.

4. Drive Metal Plate Under Leaking Shingle.

Fig. 47. Replacing asphalt shingles and repairing leaks.

Seriously damaged shingles must be removed and replaced. Insert a hacksaw blade under the shingle and cut the nails holding it to the roof. Pull out the shingle and repair any holes or rips in the roofing felt with a patch and some roofing cement. Install a new shingle and nail it in place. Use roofing cement on the bottom of the shingle (Fig. 47).

REROOFING

A good roof should last quite a number of years, but eventually you will be faced with the problem of laying a new roof. This can be done by completely removing the old one and laying a new one, or reroofing over the old material. Reroofing is possible if the old roof consists of a *single* layer of asphalt shingles, roll roofing, or wood shingles.

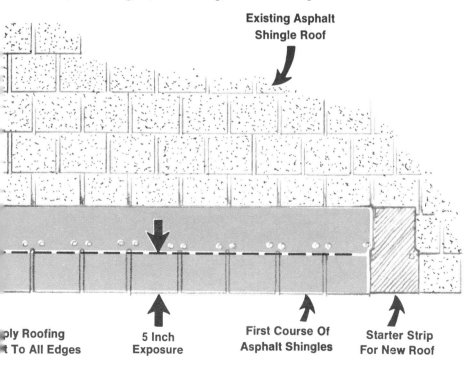

Fig. 48. Applying asphalt shingles over an existing asphalt shingle roof.

You should never reroof if the sheathing shows signs of damage or rot. You can check for this by examining the bottom of the sheathing in the attic or attic crawl space. If these conditions exist, remove the old roofing materials, repair the sheathing, and lay a new roof.

When reroofing with asphalt shingles over an existing roof of the *same* material, begin with a starter course of shingles along the roof eave. Lay a second course of shingles over the first one to form a double row, staggering the breaks in the shingles (Fig. 48). Lay each course toward the ridge with about a 5 inch exposure for the shingles. The nailing method is identical to the one described in the section covering asphalt shingling.

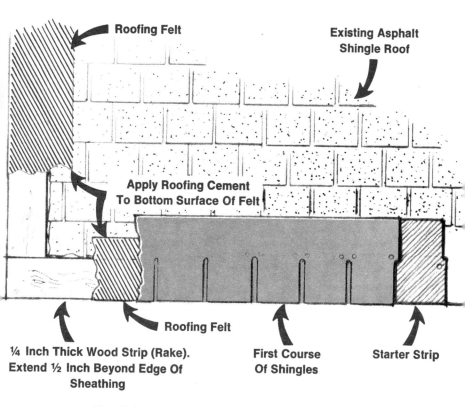

Fig. 49. Alternate method of reroofing with asphalt shingles.

There are also a number of special roofing compounds available in different colors that can be used to cover an old shingle roof. Your local dealer will have information about them and instructions on how to apply them.

Labor costs and the time required to do the job are significantly reduced if the old shingles (or other roof covering materials) do not have to be removed. Furthermore, the additional thickness of roofing material reduces the amount of heat loss in the winter. Finally, if the work should be interrupted by rain, the interior of the structure is protected by the old roof.

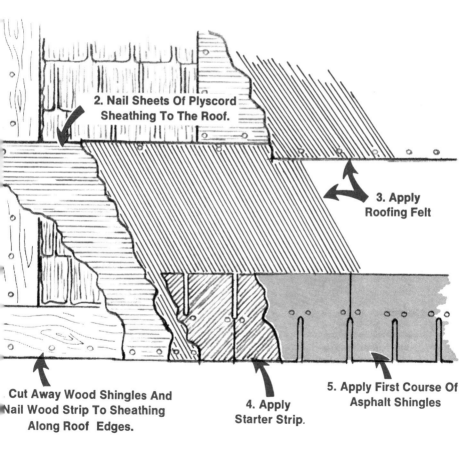

Fig. 50. Covering wood shingles with asphalt shingles.

Asphalt shingles can also be applied over wood shingles. In order to do this, the wood shingles should be in reasonably good condition and *dry*. Replace rusty or loose nails with new ones, and cover the heads with roofing cement. Before reroofing, nail down loose shingles and repair any leaks you find.

Some roofers prefer to remove the old wood shingles along the eave and gable, and replace them with 1 × 4 (or 1 × 6) boards nailed to the rafters. The roof is then covered with plyscord sheathing which is nailed to these boards and to the rafters under the wood shingles with nails of sufficient length. Roofing felt, flashing and a starter strip of asphalt shingles is then applied as previously described. (Fig. 50)

GUTTERS AND DOWNSPOUTS

Gutters and downspouts (leaders) are generally made from metal or plastic, although wood gutters are sometimes encountered on older houses and buildings (Fig. 51).

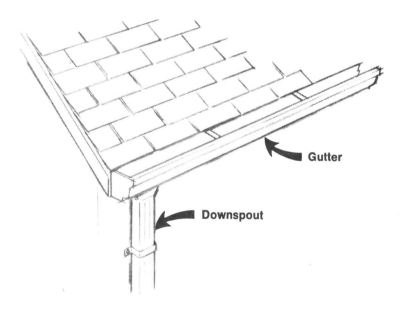

Fig. 51. Gutters and downspouts.

There are two methods of attaching gutters. One method is to nail or screw the gutters to the roof sheathing by means of a flange or strap after which the shingles or other roof covering materials are applied. Another method is to nail or screw the gutter to the fascia of the structure after roofing has been completed (Fig. 52).

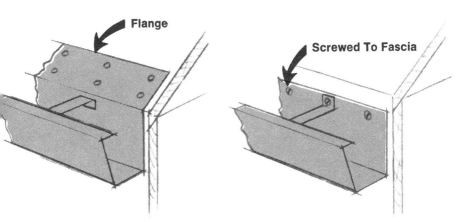

Fig. 52. Methods of attaching gutters.

Aluminum gutters and downspouts are recommended for replacing existing ones, because they are light, strong and seldom subject to corrosion. When they do corrode, it is generally in a locale near salt water.

Gutters can become clogged with leaves or other debris. This can result in water overflowing from the gutters and running down the walls, resulting in possible interior wall damage. The gutters can be kept free of debris by covering them with screens. The screens must be cut wide enough to extend up under the shingles or other roof covering materials, and to bend over the outside edge of the gutter. Make sure the gutter screen also covers the opening to the downspout (Fig. 53).

Clogged downspouts can be flushed out with a garden hose or cleaned out with a plumber's auger (Fig. 54). If neither of these methods work, the downspout should be replaced with a new one, preferably one made of aluminum.

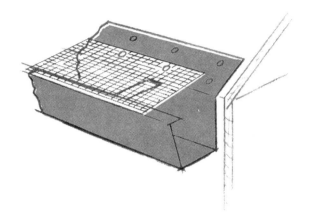

Fig. 53. Gutter screens.

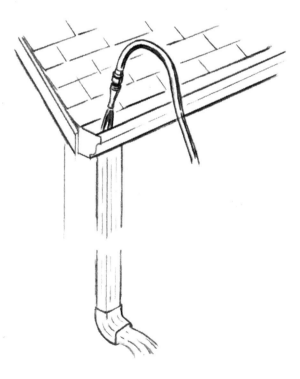

Fig. 54. Flushing out clogged downspouts.

If water collects in the gutter and does not drain toward the downspout, then the gutter does not have enough slope to it. Detach the gutter and reattach it so that there is at least a slight downward pitch toward the downspout. Before reattaching the gutter, fill in the old nail or screw holes with roofing cement. The heads of all nails or screws used to attach gutters should be covered with roofing cement. If the gutter strap or flange is fastened to the roof sheathing rather than the fascia, it should be fastened *under* the roofing felt.

Small holes in gutters can be repaired with a sealer, but it must be one that will bond with the gutter material. For example, aluminum caulking is recommended for aluminum gutters. Galvanized gutters on the other hand, should be soldered. A suitable epoxy cement or glue can be used successfully on a number of different materials. Your local building supply dealer can recommend a suitable sealer.

ROLL ROOFING

Roll roofing is an asphalt coated material used both as a roof covering or as an underlying base for tile, slate, and other roofing materials. Both smooth surfaced and mineral surfaced roll roofing are available. The latter is recommended if the roll roofing is to be used as roof covering, because the mineral granules protect the asphalt coating of the material from weather conditions.

Smooth surfaced roll roofing is often used on farm outbuildings, sheds, or temporary structures. It is inexpensive, easy to install, but rather unattractive. Mineral-surfaced roll roofing is preferred when appearance is an important factor, and is available in a number of different colors or color combinations.

Follow the manufacturer's instructions when applying roll roofing. It is generally available in 36-inch wide strips, and should be applied with each course overlapping the preceding one by approximately half the width of the roll. Roll roofing is also available with the lower half coated with mineral granules and the upper half smooth. With this type of roll roofing, the smooth portion is overlapped by the next highest course.

The manufacturer of the roll roofing will generally provide instruction on how to apply it, and will recommend the type and spacing of nails to use.

Some roofers use a starter strip laid along the roof eave. The starter strip should be at least one half the width of a regular roll roofing course (i.e. 18 inches). If it is mineral surfaced roll roofing, then the mineral surfaced slide should be laid against the sheathing.

Whenever applying roll roofing, cover the area over which the material is to be laid with asphalt roofing cement. As you lay each course, roll it smooth with a roofer's roller to eliminate air pockets and wrinkles, and to insure a tight bond. After each course has been rolled, nail it to the sheathing every foot or so. The nails are placed in the upper half of each course (i.e. the portion to be overlapped by the next highest course) (Fig. 55).

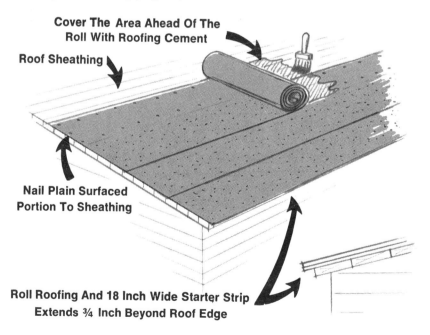

Fig. 55. Applying roll roofing.

ROLL ROOFING REPAIRS

Small holes, rips, or loose seams that are not too extensive can be repaired by applying roofing cement and pressing the roll roofing back in place. Larger holes or rips can be

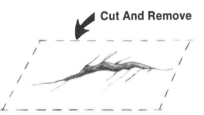

1. Cut Away The Roll Roofing Around The Damaged Portion And Brush Clean The Exposed Surface

2. Cover The Exposed Area And The Surrounding Roll Roofing With Roofing Cement.

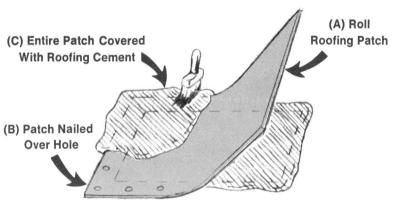

3. Cut A Patch Several Inches Larger Than The Hole In The Roll Roofing (A) Nail It Over The Hole (B), And Cover It With Roofing Cement (C).

Fig. 56. Repairing roll roofing.

repaired as illustrated in Fig. 56. The procedure can be outlined as follows:

1. Cover the damaged area with a piece of roll roofing cut several inches larger.

2. Coat the bottom of the roll roofing patch with roofing cement.
3. Nail the patch in place with roofing nails (large head types).
4. Apply roofing cement to the top and edges of the patch to seal it.

SECTION 4

Built-Up Roofing

Built-up roofing is installed on flat roofs or roofs with a very low pitch. This type of roof covering consists of several layers of lapped roofing felt and alternating layers of asphalt roofing cement, or hot roofing pitch. Applying built-up roofing is difficult and requires special equipment. Only those with experience should try to apply built-up roofing.

The first layer of roofing felt is laid along the roof eave or edge and nailed to the sheathing. The nails should be positioned approximately 5 inches apart and about 1 inch from the edge of the strip (Fig. 57). Each strip of roofing felt is 36 inches wide. Measure in one third the width and strike a line parallel to the edge of the strip. Nail along these two lines so that the nail heads are one foot apart and staggered (Fig. 57).

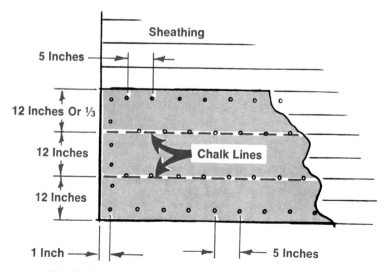

Fig. 57. Applying first layer of roofing felt on a built-up roof.

Cover the first layer of roofing felt with asphalt roofing cement, and immediately cover it with a second course of roofing felt. The second course should overlap the first one by at least 6 inches, but not more than 10 inches. Nail the *bottom* edge and then roll the second course so that it is securely bonded with the first course and all air pockets and wrinkles have been eliminated. Finish nailing the second course in the same way that you nail the first one.

Continue laying courses of roofing felt until the entire roof is covered. Use the same procedure described for the first and second courses (Fig. 58).

When you have completely covered the roof with roofing felt, cover the entire surface with a layer of asphalt roofing cement. Greater resistance to weather conditions can be obtained by sprinkling gravel or some other suitable mineral before the final coat of roofing cement dries. If you finish

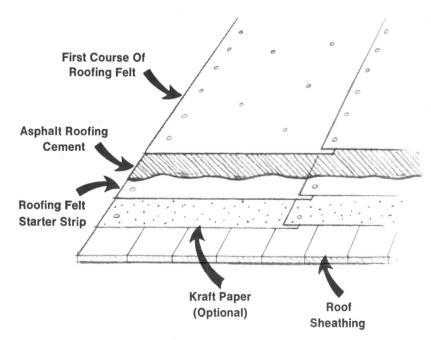

Fig. 58. Applying various layers of a built-up roof. Each layer of roofing felt is covered with hot asphalt cement. The number of layers of roofing felt will depend on job specifications.

with gravel, a retainer strip should be nailed along the roof edge (Fig. 59).

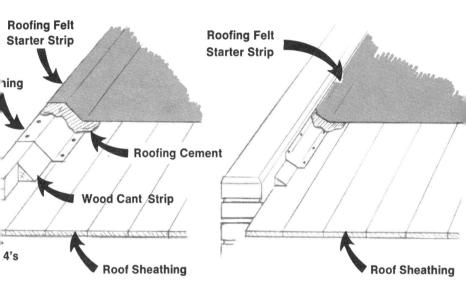

Fig. 59. Gravel retainer strips.

REPAIRING BUILT-UP ROOFING

Leaks in built-up roofing usually result from blisters or cracks that develop on the surface. Both blisters and cracks should be cleaned first before attempting to repair them. Brush any gravel, dirt or dust out of the damaged area. If the damaged area is small enough, it can be covered with roofing cement. Larger areas should be repaired with a patch cut from roofing felt.

SECTION 5

Wood Shingling

Wood shingles are generally available in lengths cut to 16, 18, and 24 inches. The widths will vary on a random basis. Shakes are similar to wood shingles in appearance, but are cut thicker and larger.

Wood shingles are most commonly used on roofs with a ⅓ pitch (i.e. 4 inches rise for each 12 inches or run) or more.

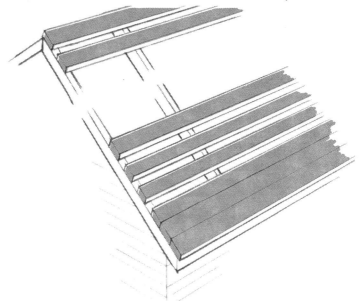

Fig. 60. Sheathing spacing for wood shingle roofs. In this example, the three boards along the eave are not separated by a space. Some roofers prefer to space all the sheathing boards. After the sheathing is nailed in place, it is covered with roofing felt.

On roofs with less than ⅙ pitch, a professional roofer should do the work because special application procedures are involved.

Many workers lay wood shingles directly over the roof sheathing on new roofs. Wood shingles tend to retain moisture. If there is no roofing felt between the sheathing and the shingles, the extra ventilation will enable them to dry out faster.

Even better ventilation is provided by laying the shingles on wood slats (shingle lathe). The slats consist of nominal 1 × 3 or 1 × 4 boards nailed to the roof rafters and spaced an equal distance apart (Fig. 60). Three slats joined together, or a board of approximately equal size, is nailed along the roof eave when wood shingles are used (Fig. 60).

Wood shingles are nailed to the sheathing or shingle lathe with zinc coated, corrosion resistant nails. At least two nails should be used to fasten each shingle. They should be placed at least 1 inch from any shingle edge, and at least 1 inch under the overlay of the next course.

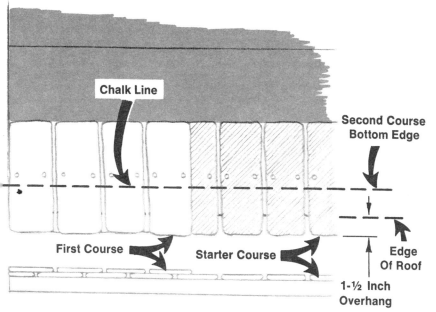

Fig. 61. Laying the starter course and first course of wood shingles.

Because wood shingles absorb moisture, they should be spaced approximately ¼ inch apart to allow for expansion. If you do not allow for the expansion, the shingles will buckle and split.

Each shingle should be attached to the roof with two nails spaced 1 inch from the edge of the shingle and 1 inch above the bottom edge of the next course. Mark this bottom edge with a chalk line as shown in Fig. 61.

The procedure for applying wood shingles to a new roof may be summarized as follows:

1. Cover the sheathing or shingle lathe with asphalt roofing felt and lay the flashing in the valleys.
2. Nail a starter course of shingles spaced ¼ inch apart along the roof eave. Allow the starter course to overhang the roof edge 1½ inches.
3. Cut a shingle to fit the angle of each valley and nail it to the roof (Fig. 62).

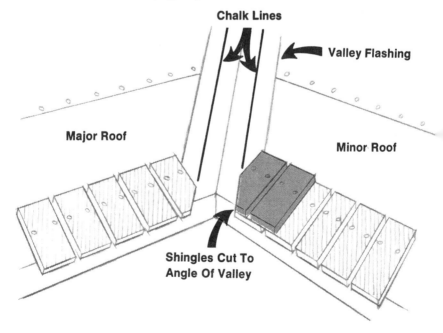

Fig. 62. Cutting wood shingles to fit valley.

4. Lay the first course of wood shingles over the starter course so that the vertical joints do not align with those below (Fig. 63).
5. Stagger the vertical joints in each course as you lay the shingles up the slope of the roof.

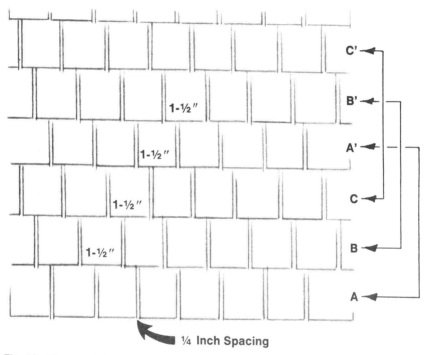

Fig. 63. Alignment of vertical joints. Joints of every fourth course of wood shingles should be alignments. Shingles should be spaced ¼ inch apart. Side lapping of shingles should be at least 1½ inches.

REPAIRING WOOD SHINGLES

The expansion and contraction of wood shingles sometimes causes them to crack and split. This tends to be one of the principal problems with wood shingle roofs, but it is not a difficult one to correct. Cut a piece of copper, aluminum, tin or roofing felt to size and insert it under the damaged shingle. Nail both halves of the shingle to the roof, and cover each nail head with a dab of roofing cement (Fig. 64). Galvanized roofing nails are recommended for nailing the shingle, because they resist corrosion well. Warped shingles should be split lengthwise with a chisel, and repaired in the same way.

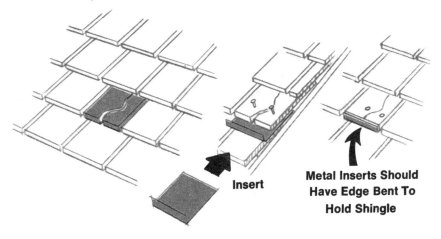

Fig. 64. Repairing wood shingles

Badly damaged or rotten wood shingles should be replaced. Remove the unwanted shingle by slipping a hacksaw blade under it and cutting the nails. The new shingle should match the color of the old one as closely as possible.

Never paint wood shingles. Shingles can be stained if you want to change or add color, but this should be done before they are laid.

SECTION 6

Tile Roofing

Roofing tiles are made from a variety of different materials, including shale, shale and clay, cement, cement and asbestos, and metal. Both curved and flat types are available for installation on roofs. Curved tiles are manufactured in the mission and Spanish style designs; flat tiles in the shingle and interlocking designs.

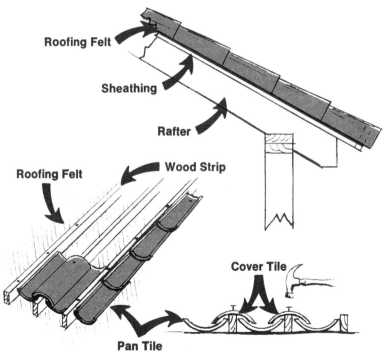

Fig. 65. Installing mission tile. Roofing felt is first laid over the sheathing and then wood strips are nailed down at spaced intervals.

Mission roofing tiles are laid in courses which overlap on alternating sides. The concave side, or cover, forms a course

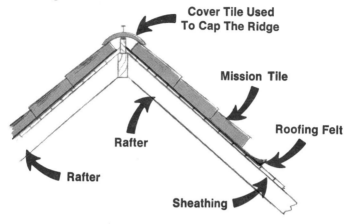

Fig. 66. Using a cover tile to cap the ridge.

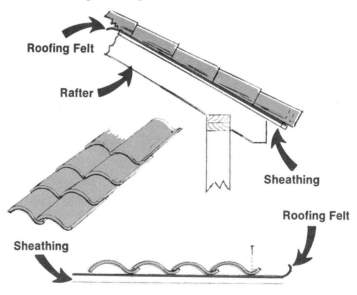

Fig. 67. Installing Spanish tile.

of tiles which overlaps with the convex side, or *pan,* of two parallel courses on either side (Fig. 65). The cover tile is nailed to a wood strip fastened to the roof sheathing; whereas the pan tile is nailed directly to the sheathing. The roof ridge is capped by a cover tile which is nailed to a wood strip fastened to the ridge (Fig. 66).

Spanish roofing tiles have a flange or lip running the length of one side of the tile (Fig. 67). The tiles are fastened to the roof with nails through the flange. The roof ridge is capped with a tile similar in design to the mission-style cover tile.

Shingle tiles are flat tiles that are nailed to the roof in much the same way that wood shingles or shakes are. The roof ridge is capped with a curved tile resembling a mission-style cover tile (Fig. 68), which is nailed and cemented in place.

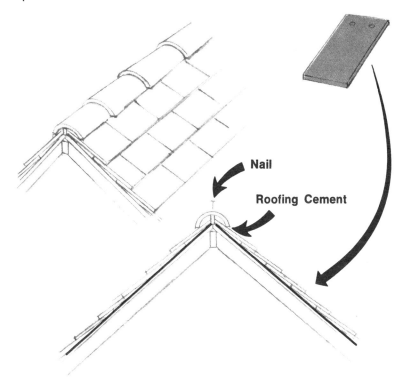

Fig. 68. Installing flat shingle tiles.

Another type of shingle tile is the one illustrated in Fig. 69. These tiles are held in place by a lip or leg at one end which hooks over a 1 × 3 strip of wood nailed to the sheathing.

Interlocking flat tiles have a bend or lip at each end which interlocks with the courses above and below it. The ridge tiles are also flat and are cemented together (and to the ridge) to form an angular cap.

Always examine the roof framing before laying a tile roof. This is a very heavy roofing material, and it requires strong framing to support it. It may be necessary to add bracing to carry the additional weight. The bracing should consist of 2 × 4's cut to fit between the sheathing and the floor, and placed every five feet or so on either side of the roof ridge.

Tile roofs are very slippery and you should be extremely careful when walking on this type of surface. Watch your

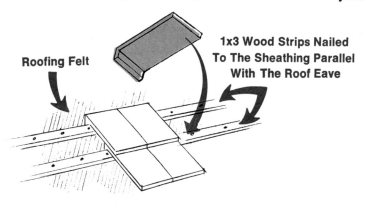

Fig. 69. Installing shingle tile with single lip.

footing, and use a safety rope and harness. Nonmetal tiles are also brittle, and you should take care to distribute your weight evenly when walking on them. Try to step on two tiles at a time rather than placing your weight on a single tile.

REPAIRING A TILE ROOF

A cracked tile can be repaired with a number of different synthetic sealers. Ask your local building supply dealer for advice. If the tile is too damaged to repair, remove it and replace it with a new one. You may have to cut nails holding the broken tile to the roof. This can be done with a hacksaw blade or nail ripper inserted under the tile. Examine the surface of the roofing felt under the tile for rips or holes. These should be repaired with a patch and some roofing cement before inserting the new tile.

Cut a piece of copper or aluminum flashing to fit under the new tile. The flashing should be long enough and wide enough to lie under the adjoining tiles. It should also be about ¼ inch longer than the length of the tile when installed. The extra ¼ inch is bent up to hold the new tile in place. Nail the flashing in place and cover the nail heads with roofing cement.

SECTION 7

Slate Roofing

Slate is a heavy, durable, nonporous rock material that formerly enjoyed considerable popularity as a roof covering. It is difficult to work with, not only from the standpoint of the material, but also because it is slippery to stand on and can crack under the weight of a man. The same precautions taken with tile roofing also apply here.

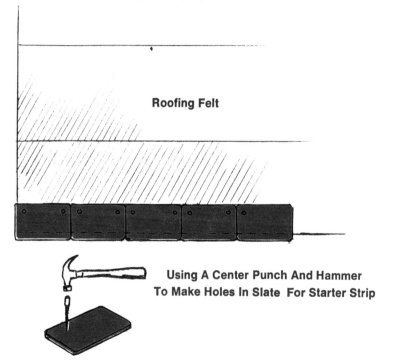

Fig. 70. Laying a starter strip of slate.

Because slate is a heavy roofing material, check the roof framing to be sure that it is capable of supporting the additional weight. The instructions for bracing a tile roof will also apply here.

Begin work by covering the sheathing with overlapping layers of roofing felt. Let each layer overlap at least 2 to 3 inches and staple it to the sheathing.

Lay a starter course of slate along the edge of the roof, allowing the slate to extend past the roof edge approximately 1 inch (Fig. 70). The slate in the starter course is laid *lengthwise* along the edge of the roof. Nail holes are punched

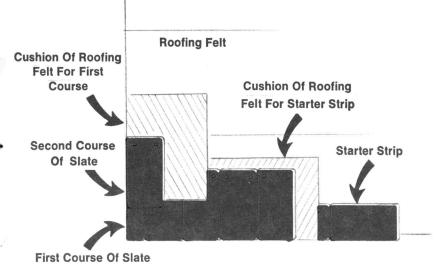

Fig. 71. Applying slate to the roof.

in the slate with a center punch and hammer. Copper wire nails *at least* 1½ inches long are used to attach the slate to the sheathing.

Some roofers prefer to lay a strip of roofing felt over the starter course and staple it to the sheathing. This will serve as a cushion between the starter course and the first course of slate.

The first course of slate is laid over the starter course, but with the *width* of each slate paralleling the eave. The first course of slate should be started with a half piece so that the vertical joints are staggered for each course. Cut the first slate by scoring it deeply down the middle with a chisel and breaking off the unwanted half by tapping it with a hammer. On most roofs, the courses should overlap about 3 inches. A 2 inch overlap is permitted on steep roofs. (Fig. 71).

Do not nail the slate down too tightly or you may cause the slate to split. Drive the nail in so that it is flush with the slate. Coat the edge of the last course of slate along the ridge of the roof with roofing cement.

REPAIRING A SLATE ROOF

A cracked slate can be repaired by filling the crack with asphalt roofing cement, putty or a suitable synthetic sealer. These materials can also be used to reattach a loose slate.

A slate too damaged to repair should be removed and replaced with a new one. If nails are still holding a portion of the slate to the roof, cut them with a hacksaw blade or nail ripper inserted between the roof and the slate. Cut a new piece of slate to fill the space. Slate can be trimmed to size by scoring deeply along a line with a chisel and breaking off the excess by tapping lightly with a hammer. Hold the piece of slate on a smooth surface, and punch new nail holes with a center punch. Do not allow the new holes to line up with the old ones on the roof. Fill in the old holes with roofing cement.

Nail a strip of metal flashing to the roof. It should be long enough so that the upper portion extends under the slate above the replacement piece, and its lower portion is about ¼ inch longer than the new piece of slate. Nail the new

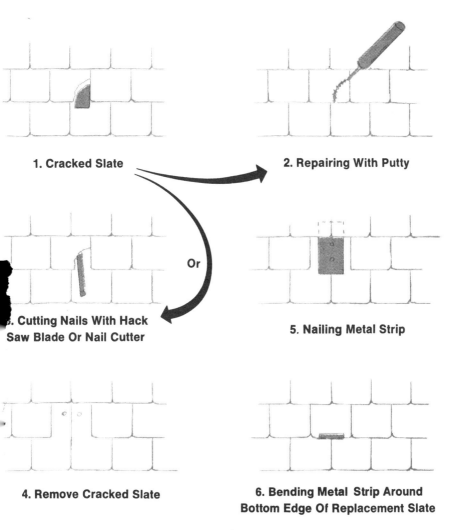

Fig. 72. Repairing and replacing slate.

piece of slate down, and bend the lower extended edge of flashing up to hold it in place (Fig. 72).

SECTION 8

Metal Roofing

Different types of metals are used as roof coverings. These roofing materials are available in thin, lightweight sheets of galvanized iron, aluminum, copper, roofing tin (terne), zinc, and monel metal. Corrugated or crimped sheet metal roofing is the easiest type to apply (Fig. 73).

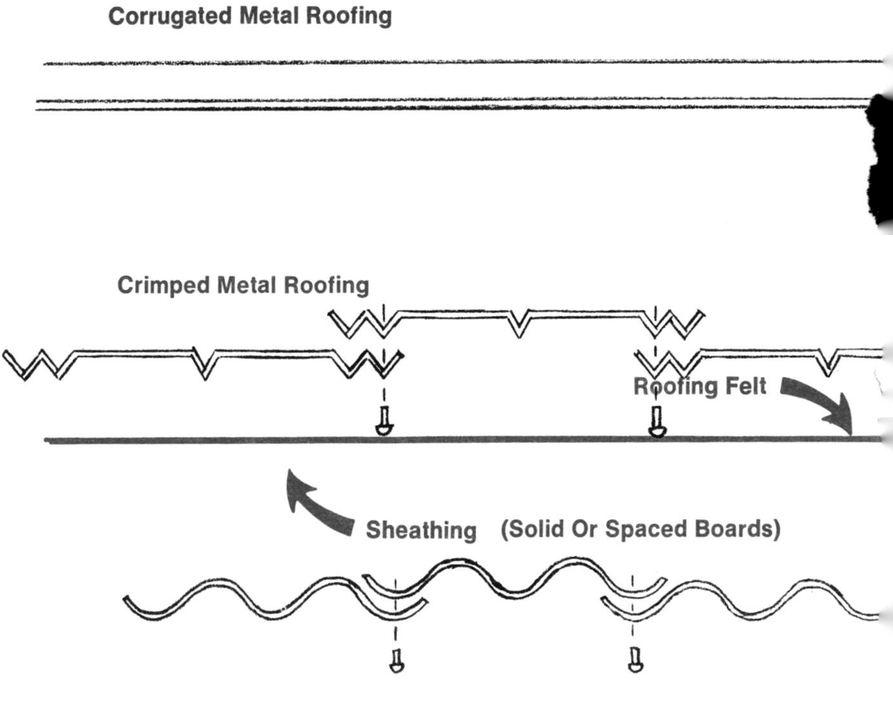

Corrugated Metal Roofing

Crimped Metal Roofing

Roofing Felt

Sheathing (Solid Or Spaced Boards)

Fig. 73. Metal roofing.

The manufacturer of a metal roofing material will generally provide detailed instructions about installing it. Read these instructions carefully and be sure you understand them *before* you begin roofing. If you have any questions, the manufacturer's local representative should be able to answer them.

Always use nails of the same material as the metal roofing. Different metals sometimes react to one another and corrode. The manufacturer of the metal roofing will specify the type of nail to use in the instructions. If instructions are not available, ask for advice at your local building-supply house.

Metal roofing can be applied over an existing roof, but only after the old roofing has been properly prepared. It is essential that all loose shingles be nailed down; damaged ones replaced; and any leaks located and repaired. Added protection can be obtained by covering the old roof with roofing felt before laying the metal roofing.

REPAIRING METAL ROOFS

Holes and broken seams are the most common source of leaks on metal roofs. Other sources of leaks include loose nails, faulty joints, and improper installation of the metal sheets. If the metal sheets have not been installed properly, they should be removed and relaid. Loose nails should be replaced. A faulty joint can sometimes be soldered.

Small holes can be sealed by filling them with roofing compound, a drop of solder, putty, or a suitable synthetic sealer. Clean the area around the hole with steel wool before filling it. A large hole will require a patch *of the same material*. The area to be covered with the patch should be rubbed with steel wool or emery cloth until the metal is bright. A metal brush is recommended for this purpose if the roofing metal has been painted. Apply an acid flux to the area around the hole and to the bottom of the patch. Tin each surface with a thin coat of solder. Soldering a patch to a metal roof is difficult, and should be done by an experienced worker.

A painted metal roof occasionally requires repainting. Although there are paints that can be applied over rust, it is a good idea to remove any rust and loose paint with a stiff wire brush before repainting.

Consult your local paint dealer before repainting. Different paints are required for the different types of metal used in roofing. A primer coat may be necessary. Never paint copper roofs. It is possible to paint colored aluminum roofs, but a special exterior paint is required for the job. Ask your local dealer about this.